FUN AND GAMES
WITH MATH

By

OUIDA SIMMONS

ISBN: 1-4107-9494-6 (e-book)
ISBN: 1-4107-9493-8 (Paperback)

This book is printed on acid free paper.

1stBooks - rev. 11/13/03

DEDICATION

This book is dedicated in loving memory to my two sons, Larry Joe Simmons and Thaddeus "Thad" D. Simmons. They both died from Heart attacks and complications of the same. A percentage of all sales from this book will go to the two boys, Hunter Simmons age 11 and Logan Simmons age7, left behind by Thad and also to The American Heart Association for research.

INTRODUCTION

It is not how well students are taught to add, subtract, multiply, and divide that counts but how well they are taught to think logically. All students will not be mathematicians, scientists, doctors, lawyers, or teachers, but all will be consumers of one kind or another. All need to know how to judge a bargain. They need to be prepared to live in the competitive world of tomorrow where calculators and computers will be doing more and more of the computations. Logical thinking will help them program these machines to calculate for them.

This book will show teachers and parents how to make math an active, living, integrated part of daily life. Math will then become more enjoyable and less of a drudgery for the students as they prepare for the world of tomorrow—today.

TABLE OF CONTENTS

MATH: A GAMING SITUATION

I have used many games to enrich, supplement, and reinforce math skills at all elementary grade levels. The simple game of BINGO can be used to teach many math concepts. With first graders, recognition of numbers is enhanced by the game. Perceptual skills are strengthened by the child having to look across for the letter and down for the number called. Place value is another concept that can be practiced with this game. To do that, when calling the numbers, say "O—six tens and four ones" or "I—one ten and nine ones."

CHECKERS

This game can be used with children ages 6-12 and beyond. Most of my students know the basic moves of the game; if some do not, they can easily be taught the moves and objective of the game. Then drill can begin on any facts or concepts that need to be practiced.

JIGSAW PUZZLES

A deck of cards can be made with problems that need to be practiced. These can be put on index cards or pieces of paper. The cards are then stacked on the table. Commercial puzzle pieces are scattered on the other side of the table. Before a player can take a piece of the puzzle, he must choose a card with a problem and answer it. Then the puzzle piece can be inserted where it fits in the puzzle. Two to four people can play the game. The same cards that were made for the checker game can be used for this game.

A dart ball game can teach many math concepts subtly. Each player gets to throw three velcro balls at a felt cloth bull's eye. Then the players add the three partial scores to get a total score. After all players have thrown their three balls and added their scores, more learning takes place. Ordinality is emphasized when individual scores are ranked first, second, third, etc. Randomly comparing two scores at a time reinforces the concept of $>$ $<$ (greater than and less than)

I remove all face cards from a deck of playing cards and then shuffle the ones remaining. The cards are placed on the table face down. From two to four players can play the game. A player draws two cards from the stack and either adds, subtracts, multiplies, or divides the numbers appearing on the cards. For instance, if the four of diamonds and the five of spades were drawn, the player might say, "4+5=9 or 5-4=1 or 4x5=20 or 5-:- 4=1 r 1." If a player can- not answer the problem correctly, the two cards are put back in the stack of cards to be drawn again by another player. If the player answers the problem correctly, the child gets to keep the cards until the end of the game. The winner is the one with the most cards at the end of the game.

I have a pair of foam rubber dice that I use most of the time for this game. I also use dice made from construction paper as shown in the illustration that follows. Any combination of numbers of dots can be drawn on the six sides of the dice. It is best to have fewer dots for primary children and more dots for intermediate students.

I use this game for speed drills in addition and multiplication. Two players can play, or one can play alone. The object of the game is to get three dominoes in a row either vertically, diagonally, or horizontally. A set of dominoes is placed on the playing surface with the dots face down. I use two desks pushed together for the playing surface, but a table can be used. The players each have cards with combinations of numbers placed randomly on them.

This is a game where bean bags are tossed at a grid with revolving X's and O's on the nine squares that make the game-board. When a square is hit with the bean bag, it tilts over to expose an X or an O. To use this game to teach math, numbers are substituted and taped on for the X's and the O's.

Game-boards encourage the child to use strategy in order to win. They give subtle practice painlessly in skills needed.

These teaching aids are designed to increase speed in drills. Addition, subtraction, multiplication, and division can be practiced on tachistoscopes. It depends upon what process needs to be studied. The illustrations that follow are just examples of what can be done with these devices. I am sure you can think of other shapes equally attractive that will interest children. With older students, the shape of a car might be more appropriate. These learning aids can be used by one child or used to instruct a group of children.

These two games are played like BINGO with cards and cover pieces. Cards can be made with addition, subtraction, multiplication, and division facts on them. Flash cards can be purchased at most drug stores or department stores for less than $2 per box and serve equally as well as those made. Playing cards can be made from poster board cut in nine inch squares. Write the words COMBO at the top front of the squares and write FACTO on the top back of the same card. Mark off squares for individual numbers to be placed at intervals in all spaces, with the word FREE written in the center like the examples below.

Do not throw away old calendar pages. They can be laminated and used as a playing board for a math game. At the primary level, recognition of numbers can be reinforced with a game I call CALENDO. The game is played similar to BINGO. If a child covers a row of numbers horizontally or vertically, he wins. Questions can be made according to the needs of the children.

This is another game played like BINGO. Geometric shapes are used rather than numbers on the playing cards. Flash cards with geometric shapes on the front and the name of the shape on the back are made to play this game. The cards are shown to the players (at first the shapes—later the names). If the player has the shape, it is covered. The first player to get five shapes in a row covered is the winner. Following are some illustrations of the flash cards and a playing card made from poster board.

Chalkboards are inexpensive to make. Masonite or plywood sawed into a rectangular piece about 10"x12" can be painted with chalkboard paint. Old tube socks make ideal erasers and can be washed as often as needed.

Missing addends in subtraction are easy if you just think "add." With the problem 18 - 9 =? I say to the students, "Use your head as a calculator and punch the 9 into the calculators memory, and it will remember the number 9 for you. Now count to eighteen. You find your answer is nine. You really think 9 +? = 18. As long as the child needs to, this counting can be done on the fingers like this: say, "9 (punch it in memory)-10-11-12-13-14-15-16-17-18." Each time a number is said, a finger goes up. When the number eighteen is reached, you read the number of fingers you raised as you counted the answer on your hand. Of course the answer or the number of fingers is 9.

To help the child remember better, the plus symbol can be made of sandpaper. Let the child feel the shape as he looks at it and says' "PLUS means put together." This use of sight, sound, and the tactile senses gives the child a multisensory approach to the understanding of these abstract symbols. The minus and equal symbols can also be made from sandpaper and used the same way.

Place value is essential for understanding other math concepts. Therefore, it needs to be taught early. I start with concrete objects like counting sticks (pop-sicle sticks are good). Have the children separate the sticks into groups of tens and ones. I walk around the room and select several children to show me their sets of tens and ones. Then I go to the board and write the numeral that stands for the individual groups of tens and ones I see on their desk. Next I ask the students to show me a numeral like: 15 or 1 ten, and 5 ones. We continue this activity until they become proficient with the exercise.

Digital clocks and watches have almost made reading the traditional clock obsolete. However, it is still in math textbooks, so we must teach it. I use individualized clock faces made from poster board with the hour hand colored a bright orange and the minute hand brown or another color. I show the students that the number my mouth says first is the hour and the numbers I say next are the minutes as I write (5: 30) in standard notation on the chalkboard. I ask students to place the orange hour hand on the five then move the brown minute hand slowly from its home position on the twelve to the right. As the hand passes a number, count 5-10-15-20-25-30. Now your clock is showing (5: 30), with the hour hand on the five and the minute hand on the six.

In the beginning each student is given a packet of play money. The money can be bought from a school supply store or most variety drug stores. Most consumable math books have these manipulatives in the back of the texts. My source is from the textbooks. We first learn the value of pennies, nickels, dimes, and quarters in that order; then we learn the paper bills.

We learn to distinguish one coin from another by size and engravings. The color of the copper penny makes it easy to find.

Most any subject can be taught by using the newspaper. In this section I tell how I have used it with math lessons.

Weather charts and maps provide opportunities to correlate geography, science, and math. Geography lessons can be designed through the use of the weather map; science lessons can include understandings of fronts, pressure systems, and weather patterns. Math lessons can include activities such as the ones that follow: One team of two or three students can prepare activities for other teams to compete in games using weather maps, charts, and questions as shown in the following example

This information can be collected over a period of a week, a month, three months, or more, and a daily or weekly graph can be made to show the information. The temperatures around the world can be collected and graphed in the same way. Students can then make up their own problems about the graphs. When children give words to a problem, they can usually solve it.

Many intermediate students are motivated to do math by using the sports scores of the local high school teams as well as the professional teams' scores. The children need the sport section of the newspaper with those statistics listed.

This activity helps reinforce the word names for numbers. Each student is given one or two pages from the newspaper to look for number words. When number words are found, they are cut out and pasted on a sheet of manila paper or notebook paper. Primary children may need some assistance in their search for number names. The objective for this assignment can be any number words or specific number words.

Older elementary students enjoy the competition of this game. When the instructor says, "Go." The groups will have five minutes to find as many number words as possible. At the end of the five minutes, one student from each group will write the words they found on the chalkboard. The group that has the most words is the winner.

There is something about cutting and pasting in the primary grades that makes a lasting impression. When children find a numeral with two or more digits, they can read them and expand them like this: (126 is 100 + 20 + 6 or 1 hundred, 2 tens, and 6 ones. The children are given two or three pages of the newspaper to find and cut out the numerals. They cut and paste the numerals they find on a sheet of notebook paper.

Reading a TV schedule is part of daily life. It is imperative that children learn to read it adequately. Schedules come in many varieties. Some are in book form and are delivered free with a subscription to cable TV, once each month. Another comes with the local Sunday newspaper and has the TV programs listed for the week. Another weekly schedule is the popular *TV GUIDE* to which one must subscribe if it to be delivered to the home. Finally, there is the daily schedule in the daily newspaper. If children learn to read one of these types well, they should be able to read most any schedule.

The students are given newspapers with the movie listings in them. Then I tell them that they can choose a movie from the (G) rating or (PG) category. They are then instructed to look for the prices each theater charges for a movie ticket. They are told they can take a friend along, which will double their expense. They must plan to take enough money to buy popcorn and a soft drink. When they have calculated the cost at several local theaters, they decide which movie would be the best bargain and tell why. They figure the amount of money it will cost for two tickets and refreshments. They also compute how much money it would cost if they could go to the matinee. Following are some of the listings from which they might choose the movie and figure the cost.

This activity is especially good with intermediate students or older children because some of them are already thinking seriously about what they want to do when they grow up. The students are given the want ad section of the newspaper. They look through the Help Wanted section and select a job in which they are interested. Then they take the salary listed, weekly or monthly, and figure how much they will receive for a year.

The purpose of this activity is to give practical experience through student involvement in finding multiples of numbers. The students should be directed to find large numbers from ads in the newspaper and prepare two boxes (shoe boxes will do). One box will contain numbers for which multiples must be found and the other box contains multiples of all numbers in the first box. For example, the number 5, used as a stimulus, could have the following multiples 5, 10, 15, 20, 25, 30, 35, 40, 45, 50, etc.

This activity aids the student in discovering through comparison the meaning of *ratio*. You will need several sets of similar pictures of different sizes, cut out, mounted, and labeled with either size or prices used like the illustration below.

This helps the student understand the stock market and how it affects our daily life.

The greater than and less than symbol take on much more meaning when I tell the students to think of them as hungry Pac Man.

These can be precut before teaching the lesson, and students can assemble them, or they could be done entirely by the instructor for a display on a bulletin board. When displayed, these shapes become a constant reminder of the geometric shape and its name. Below are some examples of the four major geo shapes that primary children are expected to know. For intermediate children, other shapes can be pictured. They can construct their own pictures and explain how they made their geodesic forms. They could be assigned certain geo shapes to use in designing a building or some other project similar to the illustrations that follow.

An iris can be made from four circles, one rectangle, two triangles, and a square piece of construction paper on which to mount the flower. Since irises come in many colors, the circles can be made from a variety of colored construction sheets. The rectangles and triangles should be green, because they make the stem and the leaves. The square pieces of paper can be a contrasting color to emphasize the petals of the flower.

The circle, square, and triangle make pretty Christmas tree ornaments. Other geo shapes do also, but I like these best. Six circles, cut the same size, are needed for the circle ornament. Five squares, cut the same size, are needed for the square ornament. Four triangles cut from the same pattern are needed to make the triangular ornament. These shapes can be cut from colored construction paper or a heavy-weight Christmas wrapping paper.

A small group or an entire class can prepare a bulletin board using geometric shapes found in a three to five-day period in the newspaper. Colored picture collages can be made from pictures cut from magazines or newspapers to show: a point, a segment, a ray, a solid, a plane, a circle, a square, a sphere, a cone, a triangle, etc.

Math problems can be written on petals cut from paper plates. Slits that fit the petals are cut in another plate so petals can be inserted and a flower made. The center and the petals can be as colorful as your imagination allows. For four players, four plates with slits are needed for the centers of the flowers, and as many petals as needed to form the four flowers should be cut.

This activity is a pleasant diversion from the story problems in the textbook. For instance, have the children draw Halloween pictures. Show the pictures to the class. Have everyone write one story problem about the picture or children can simply tell the problems. Show another picture and do the same until about five or six pictures have been shown. Continue to show a few pictures each day until all have been shown and questions have been written or asked orally about them.

Activities using grocery store items are numerous. Children can recognize and read the prices of items. They can add the number of ounces or millimeters in a designated number of canned goods. The child can look at various packages and identify the ones containing more or less bulk weight. Intermediate children can figure out the unit price of items. Such questions as these might be asked. If tomato sauce is 3/$1, or bell peppers are 4/$1, what is the cost of one can of tomato sauce and the cost of one bell pepper?

Have the child figure the cost of specific items minus coupons. If the store doubles the value of the coupon, that is another math problem to solve. The child can multiply the value of the coupon by two and then subtract the amount from the list price. I cut coupons out of the weekly and daily newspapers and current magazines to use with my classes. Most people have access to trade coupons, too. Questions like this could be asked. If bread cost $1.49 for a loaf, what will it cost if I use this coupon worth $.50? What would it cost if the coupon's value was doubled?

The child can write down the mileage at the beginning and end of each day. Then the distance the car has been driven in one day can be calculated. If this mileage is kept for a week then the child could figure how many miles the car is driven in that length of time.

Colorful folders to keep word problems in are sometimes self-motivators. Colored Duo-tang folders with pictures cut and pasted on the outside may invite a child to look inside. Folders with Disney characters, Care Bears, Aladdin, Garfield, and Super Heroes can be purchased, but it is less expensive to decorate them yourself

Problems like those in the child's math text can be individualized by substituting the names of the children in place of those in the text. Problems become much more meaningful for John and Jan when you transpose a problem from the book like this. John bought Jan a birthday card for $.50 and a present for $2.28. How much did John pay in all? Another example might be this. Jan's mother gave her $5.00 to buy these items: bread $.91, a gallon of milk $2.49. Will Jan have enough money? How much change will she get? I am sure you can think of many more problems.

I tell my students that the sentence near the end of a word problem is usually the asking sentence. It is the sentence that asks something about the facts in the story. It always ends with a (?) question mark. Certain words in that sentence tell exactly what to do with the facts presented in the story problem. We do lots of practice just looking for the asking sentence. Then we begin practice drills in finding the key words.

Use the story problem cards made with the key words highlighted in yellow for the game. The cards can be dealt out four or five face down on the desk to each player. Taking turns, each player picks a card and reads the story either orally or silently. Since the key words are highlighted; they are easily found by the child. The child looks closely at the Magic Flower

Garden and locates the process indicated by the key word and proceeds to solve the problem.

The teaching of the math vocabulary need not be drudgery. It can be done in a meaningful but interesting way. Word search games help students to recognize and familiarize themselves with the words to be learned. At first the teacher can make the word search games. This can be done by writing the words on graph paper (one letter per space), then filling in any blank square spaces with randomly selected letters of the alphabet. If you have the software and a computer, the job can be done in less than half the time required to make them with graph paper. The students can be taught to make these word search games by either of the methods described above.

Cross-word puzzles can be constructed by both students and teacher. The overhead projector is invaluable for introducing the making of this kind of puzzle. We work in small groups. The groups start by writing definitions to the words we are using in the puzzle. Then they try to shorten definitions to use for clues. If they are doing this on graph paper, they write their clues on notebook paper and use the graph paper to fit their words inside. If the appropriate software is available to use on a computer, this puzzle making is done in a jiffy by following simple instructions that come with the software.

MATH: A GAMING SITUATION

I have used many games to enrich, supplement, and reinforce math skills at all elementary grade levels. The simple game of BINGO can be used to teach many math concepts. With first graders, recognition of numbers is enhanced by the game. Perceptual skills are strengthened by the child having to look across for the letter and down for the number called. Place value is another concept that can be practiced with this game. To do that, when calling the numbers, say "O—six tens and four ones" or "I—one ten and nine ones."

Computation skills can be reinforced or practiced with BINGO. For instance, the teacher or parent can make a set of cards with combinations that add up to all the sets of numbers on the BINGO card. In like

1

manner, subtraction problems can be placed on the backs of the addition cards in order to practice that skill. Children can use scratch paper or individual chalkboards to solve the problems. This not only provides skill and drill in computation but it teaches children to listen carefully to problems being called out to them. For example, I call, "I—(29-11) or B—(7+2)." If students need drill and practice with multiplication and division, problems like these can be used. O—(8x8) or B—(64 / 8).

CHECKERS

This game can be used with children ages 6-12 and beyond. Most of my students know the basic moves of the game; if some do not, they can easily be taught the moves and objective of the game. Then drill can begin on any facts or concepts that need to be practiced.

For example, math problems can be written on index cards for any needed practice. For first and second graders, I write simple addition and subtraction problems such as (9 - 5 =?) or (4 + 5 =?). My third and fourth graders get problems like,(4x5=?) (20 - 5 =?) or (469 + 178 =?) or (647 - 178 =?). Fifth and sixth graders get problems that are more difficult. Some examples are, (7000 - 162 =?) or (6,838 + 162 =?) or (45 x 1.25 =?) or (56.25 - 1.25 =?) or (1/2 x 1/4 =?).

Before a player can move a checker, he must first draw a card from the stack and answer the problem written on it. If the cards are laminated, a crayon can be used to figure the answer; if not, scrap paper can be used to do the computation.

JIGSAW PUZZLES

A deck of cards can be made with problems that need to be practiced. These can be put on index cards or pieces of paper. The cards are then stacked on the table. Commercial puzzle pieces are scattered on the other side of the table. Before a player can take a piece of the puzzle, he must choose a card with a problem and answer it. Then the puzzle piece can be inserted where it fits in the puzzle. Two to four people can play the game. The same cards that were made for the checker game can be used for this game.

Another variation of this game is that the puzzle pieces can be made with problems written on them. The back of the puzzle pieces can have the answer written there. The puzzle below has multiplication and division problems for the student to solve. The teacher or parent can supply the correct answers on the backs of the pieces. If drill in addition and subtraction are needed, those problems could be put on the puzzle pieces. Cut out the puzzle pieces. If these were laminated before cutting them out, students could do the computations with crayons. Cut out the puzzle pieces. Multiply or divide to solve the problems. Put together the sides with the same answers to work the puzzle. The puzzle is on the following page.

JIGSAW PUZZLE

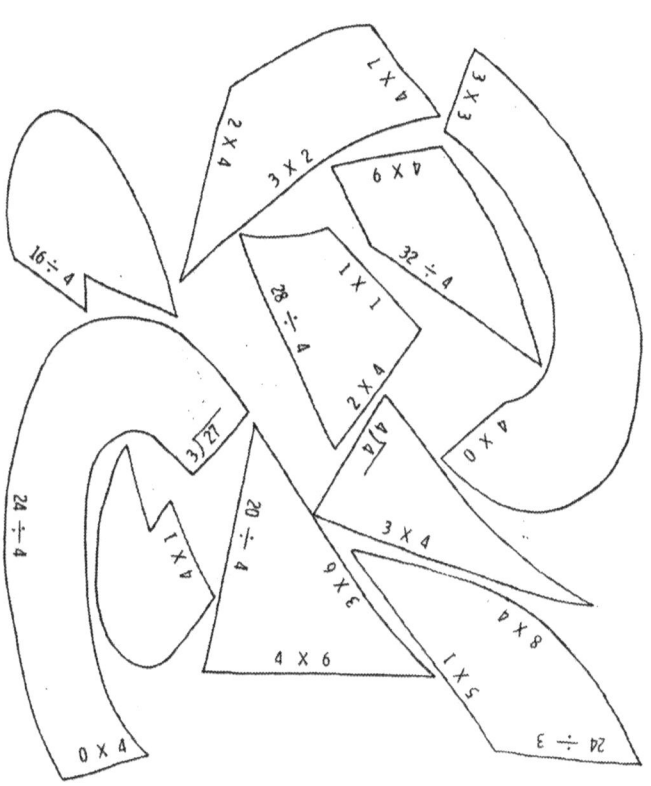

DART BALL GAMES

A dart ball game can teach many math concepts subtly. Each player gets to throw three velcro balls at a felt cloth bull's eye. Then the players add the three partial scores to get a total score. After all players have thrown their three balls and added their scores, more learning takes place. Ordinality is emphasized when individual scores are ranked first, second, third, etc. Randomly comparing two scores at a time reinforces the concept of > < (greater than and less than).

This game can be used to reinforce place value. Certain rings on the bull's eye can be designated as hundreds, tens, and ones place. I use a small piece of masking tape with the words hundreds, tens, and ones written on it. The bull's eye can be thousands place. The playing surface would look like this illustration:

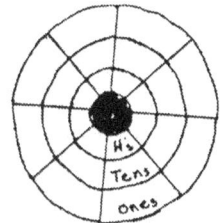   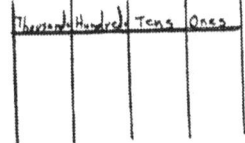

With some students, I have to make a grid on the chalkboard for the child to add the score like the illustration above and to the right. Children enjoy playing this game and do not realize they are learning math. If subtraction is being drilled, children can compare exact differences in scores by subtracting a smaller score from a larger score.

CARD GAME

I remove all face cards from a deck of playing cards and then shuffle the ones remaining. The cards are placed on the table face down. From two to four players can play the game. A player draws two cards from the stack and either adds, subtracts, multiplies, or divides the numbers appearing on the cards. For instance, if the four of diamonds and the five of spades were drawn, the player might say, "4+5=9 or 5-4=1 or 4x5=20 or 5-:- 4=1 r 1." If a player can- not answer the problem correctly, the two cards are put back in the stack of cards to be drawn again by another player. If the player answers the problem correctly, the child gets to keep the cards until the end of the game. The winner is the one with the most cards at the end of the game.

As you can see, any of the four major skills can be drilled or practiced with this game. The deck of cards and drills can also be used to play checkers or work a jigsaw puzzle as explained earlier.

DICE GAME

I have a pair of foam rubber dice that I use most of the time for this game. I also use dice made from construction paper as shown in the illustration that follows. Any combination of numbers of dots can be drawn on the six sides of the dice. It is best to have fewer dots for primary children and more dots for intermediate students.

The game can be used to practice column addition with primary students. They throw the pair of dice three times. Each time they add the number of dots on top of the dice and record the sum either on paper or the chalkboard. Then they add the three sums, from the three throws, for a grand total. Two to eight players can play the game in three rotations. The number of throws and the number of dice thrown can be adjusted to meet the needs of the group playing. First grade might throw only one die two or three times while the third grade would throw two dice four times. With my intermediate students, I have them throw three times. The children multiply the number of dots on one die by the number of dots on the other die and record the products on scrap paper or the chalkboard.

After three rotations, they add the three products for a grand total.

When totals have been tallied, we compare the various scores as to largest or smallest and use the > < symbols. This is a quiet game that can be done while others in the classroom are doing other activities.

TIC TAC TOE WITH DOMINOES

I use this game for speed drills in addition and multiplication. Two players can play, or one can play alone. The object of the game is to get three dominoes in a row either vertically, diagonally, or horizontally. A set of dominoes is placed on the playing surface with the dots face down. I use two desks pushed together for the playing surface, but a table can be used. The players each have cards with combinations of numbers placed randomly on them.

Players take turns drawing dominoes and adding the number of dots on both ends or multiplying the number of dots on one end by the number of dots on the other end of the domino. Subtraction and division can be practiced also. This is done by subtracting the number of dots on one end of the domino from the number of dots on the other end of the domino or division of one set of dots by the other set of dots on the domino. It depends on what skill needs to be practiced as to which math problems the teacher or parent uses.

Following is an illustration of a playing card that can be made for this game. Poster board is good but notebook paper can also be used. If laminated, it will last much longer. Numbers can also be changed with ease to fit the needs of students on a laminated surface.

TIC TAC PLAYING BOARD

	0	
0	0	0
0	0	1
2	3	4
4	5	6
6	8	8
9	10	12
12	15	16
18	20	24
25	30	36

TOSS ACROSS

This is a game where bean bags are tossed at a grid with revolving X's and O's on the nine squares that make the game-board. When a square is hit with the bean bag, it tilts over to expose an X or an O. To use this game to teach math, numbers are substituted and taped on for the X's and the O's.

For my young primary students we either do column addition or practice adding our total score mentally. Then we compute the sum of two or more scores. With middle primary children, we drill on place-value problems with hundreds, tens, and ones.

The score is computed by adding all digits in ones place and all digits in tens place and then all digits in hundreds place. Thus the answer would be 2 hundreds, 7 tens, 9 ones or 279.

The illustration below will show you how the TOSS ACROSS playing surface will look.

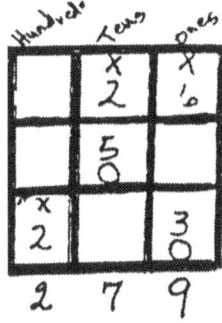

Decimal place value can be practiced by intermediate students by writing decimals numbers. These decimal numbers are taped on for easy removal.

The diagram below shows how it works.

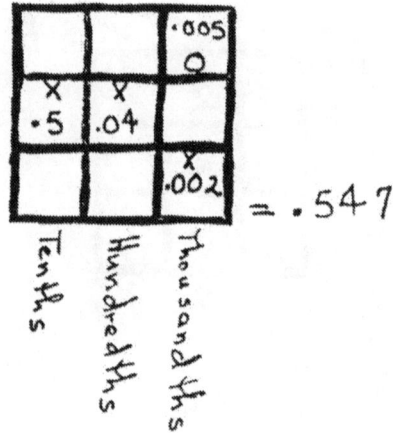

GAMEBOARDS

Game-boards encourage the child to use strategy in order to win. They give subtle practice painlessly in skills needed.

Cards are made with problems the child needs to practice. Again, index cards or notebook paper can be used for the cards. Then, before a player can move forward on the game-board, he must draw a card from the stack and answer the problem. Scratch paper can be used for the calculations. The player throws a die to see how many places to be moved on the playing board. A die can be made of construction paper and numbers written on the sides like the one illustrated on the bottom of page. Each player takes a turn drawing a problem, solving it, throwing the dice, and moving toward the goal on the game-board.

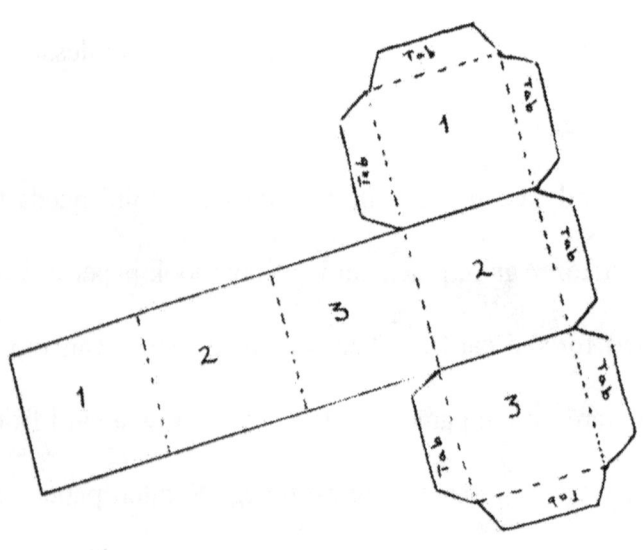

Two game-boards are included on the following pages as examples for you. I am sure you can think of many more. Or use some commercial game-boards you have.

GAME BOARD

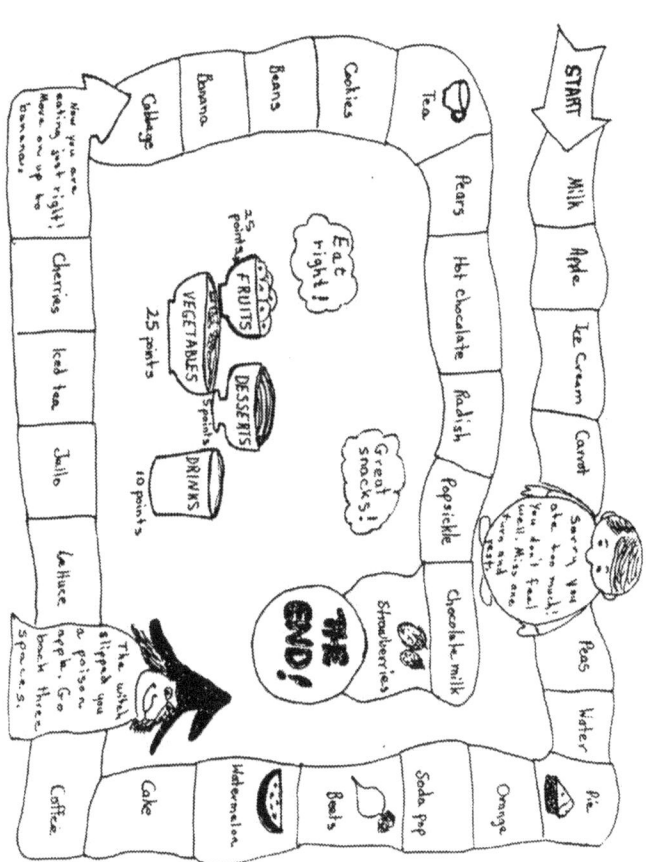

The one with the most points for eating right is the

winner.

GAME BOARD

<u>Lost Pug</u>

Help him get home!

TACHISTOSCOPES

These teaching aids are designed to increase speed in drills. Addition, subtraction, multiplication, and division can be practiced on tachistoscopes. It depends upon what process needs to be studied. The illustrations that follow are just examples of what can be done with these devices. I am sure you can think of other shapes equally attractive that will interest children. With older students, the shape of a car might be more appropriate. These learning aids can be used by one child or used to instruct a group of children.

Use poster board to construct the scopes and the strips that are pulled through them. Numbers are written on the strips so that they show through the slots. They will last longer if laminated.

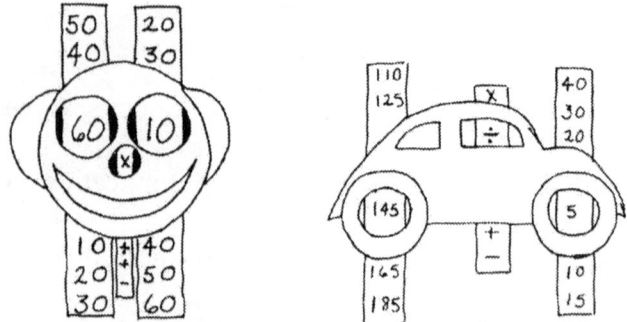

MORE GAMES YOU CAN MAKE—

COMBO AND FACTO

These two games are played like BINGO with cards and cover pieces. Cards can be made with addition, subtraction, multiplication, and division facts on them. Flash cards can be purchased at most drug stores or department stores for less than $2 per box and serve equally as well as those made. Playing cards can be made from poster board cut in nine inch squares. Write the words COMBO at the top front of the squares and write FACTO on the top back of the same card. Mark off squares for individual numbers to be placed at intervals in all spaces, with the word FREE written in the center like the examples below.

C	O	M	B	O
10	15	11	1	10
9	8	7	0	3
3	12	FREE	18	9
16	9	3	2	8
13	5	6	4	17

Addition and Subtraction

Facts

F	A	C	T	O
0	9	32	12	5
25	36	40	30	15
24	45	FREE	42	20
7	63	48	6	50
35	81	72	54	60

Multiplication and Division

Facts

Play the game in this manner. The entire class or

small groups are given COMBO or FACTO playing

cards. The teacher, or leader draws a flash card from the stack that has been placed face down on the table. The card is shown to one player who answers the problem. All players check to see if they have that answer on their game card. They cover the number if they have it. Cover pieces can be cut from scrap paper. All players have a chance to answer problems. When a player gets five numbers covered in a row, either horizontally, vertically, or diagonally, he calls "COMBO or FACTO." The winner clears his card and the leader proceeds to draw and show other flash cards as the game continues. Again, the playing cards will last longer if they are laminated.

CALENDO

Do not throw away old calendar pages. They can be laminated and used as a playing board for a math game. At the primary level, recognition of numbers can be reinforced with a game I call CALENDO. The game is played similar to BINGO. If a child covers a row of numbers horizontally or vertically, he wins. Questions can be made according to the needs of the children.

This game enhances perceptual skills. The child has to look from Sunday to Saturday in a horizontal direction to find a date such as, Tuesday the seventh. The skills of ordinality are automatically built in. If the question is, "Do you have a Tuesday the seventh?

Cover it. If your card has a Friday the thirteenth, cover it." These type questions reinforce ordinality.

It would help if the calendar pages were large enough to rank the days of the months as to first, second, third, etc. before laminating the pages for use in games.

Following are some questions that can be asked about the August calendar page that follows.

1. What day of the month is August 4th.?

   answer- Tuesday-Cover it.

2. The 19th. of August is on what day?

   answer- Wednesday—Cover it.

3. If you have a Sunday the 10th. cover it.

4. If you have a Monday the 17th. cover it.

5. If you have a Sunday the 2nd. cover it.

31

6. What day of the week is August 16th??

answer- Sunday—Cover it.

7. If you have Sunday the 9th. cover it.

8. If you have a Sunday the 23rd. cover it.

This should be a CALENDO or a win for the person holding this calendar page. Other questions can be made to ensure players will win. All twelve months can be used. Just be sure questions asked will be about the months that are used.

A calendar page follows on the next page.

SUNDAY	MONDAY	TUESDAY	WEDNESDAY	THURSDAY	FRIDAY	SATURDAY
						1
2	3	4	5	6	7	8
9	10	11	12	13	14	15
16	17	18	19	20	21	22
23	24	25	26	27	28	29
30	31					

AUGUST

SHAPO

This is another game played like BINGO. Geometric shapes are used rather than numbers on the playing cards. Flash cards with geometric shapes on the front and the name of the shape on the back are made to play this game. The cards are shown to the players (at first the shapes—later the names). If the player has the shape, it is covered. The first player to get five shapes in a row covered is the winner. Following are some illustrations of the flash cards and a playing card made from poster board.

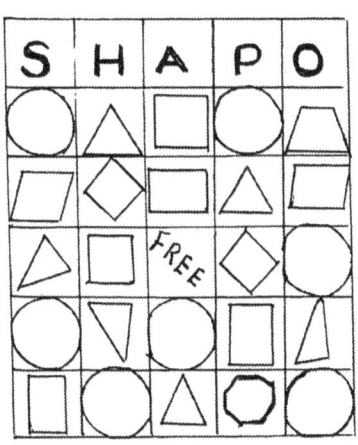

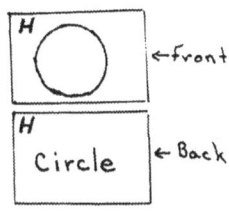

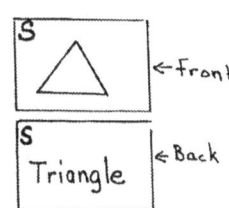

INDIVIDUALIZED CHALK-BOARDS:

A MUST

Chalkboards are inexpensive to make. Masonite or plywood sawed into a rectangular piece about 10"x12" can be painted with chalkboard paint. Old tube socks make ideal erasers and can be washed as often as needed.

Many math drills can be done by one or a group of youngsters at the same time by using individual chalkboards. Difficulties can be spotted immediately. More challenging problems can be made available to those who do not need drill while the teacher or parent works with those having trouble.

Since these drills can involve many youngsters, they motivate good self-discipline. Children have to listen attentively to the problem the teacher or parent asks them to solve. When they finish solving the problem, they turn their boards over and wait for the teacher to say, "Show me your answers." Children can take turns calling out problems to be solved, provided they have the correct answers available. The teacher can let the leader use the teacher's edition to the math text. Students like being the leader and the teacher can participate along with the class or work individually with those having difficulty.

THINK ADDING WHEN YOU

SUBTRACT

Missing addends in subtraction are easy if you just think "add." With the problem 18 - 9 =? I say to the students, "Use your head as a calculator and punch the 9 into the calculators memory, and it will remember the number 9 for you. Now count to eighteen. You find your answer is nine. You really think 9 +? = 18. As long as the child needs to, this counting can be done on the fingers like this: say, "9 (punch it in memory)-10-11-12-13-14-15-16-17-18." Each time a number is said, a finger goes up. When the number eighteen is reached, you read the number of fingers you raised as you counted the answer on your hand. Of course the answer or the number of fingers is 9.

When adding two one-digit numerals, it is easy if you again use your head like a calculator. This time punch the largest number into memory and count the other digits on your fingers. To solve the problem (9 + 8 +?), say, "9 (punched into memory) 10-11-12-13-14-15-16-17 as you count the eight fingers you are holding up. When all the fingers you held up have been counted, you have the answer. Listen to what your mouth says as you count the last finger. Yes, you said seventeen. That is your answer."

This may seem elementary, but I had a third-grade student's parents ask me to show them how to help their child with this concept. This was the same thinking process I showed them how to use with their child. It really works with children.

LITTLE + LITTLE = BIG

When we put two or more little objects together, the answer always equals something bigger.

To explain this to children, pictures of concrete objects are often necessary, like: ** + ** = ****.

To help the child remember better, the plus symbol can be made of sandpaper. Let the child feel the shape as he looks at it and says' "PLUS means put together." This use of sight, sound, and the tactile senses gives the child a multisensory approach to the understanding of these abstract symbols. The minus and equal symbols can also be made from sandpaper and used the same way.

Greater than and less than symbols can be made of sandpaper also in order to use the multisensory approach to learning them. This concept can also be shown graphically. A larger number of objects can be placed to the left and a lesser number of objects placed to the right of the sandpaper symbol to read: ******** > ****, eight is greater than four.

Another method I use with great success is pictures. The symbols for greater than and less than each look like the mouth of a hungry alligator. I tell the children he is always hungry and opens his mouth toward the largest number. I add teeth to the symbol like this: (9 > 5) or (4 < 12). I ask, "Which number is this hungry alligator going to eat?" They seldom miss the correct answer. Then we practice saying the math sentences.

RECOGNIZING TENS AND ONES

PLACE

Place value is essential for understanding other math concepts. Therefore, it needs to be taught early. I start with concrete objects like counting sticks (popsicle sticks are good). Have the children separate the sticks into groups of tens and ones. I walk around the room and select several children to show me their sets of tens and ones. Then I go to the board and write the numeral that stands for the individual groups of tens and ones I see on their desk. Next I ask the students to show me a numeral like: <u>15</u> or 1 ten, and 5 ones. We continue this activity until they become proficient with the exercise.

Using packages of dried beans is another concrete object that is inexpensive and their use reinforces the place value concept. The child can find out how many sets of tens and how many ones are left over in one half cup, a cup, or the whole package of beans. Two packages of beans can be used to find which package contains more beans. Here again the > < symbols can be used to describe the results of the discovery. You can think of other activities with these objects, I am sure.

TELLING TIME

Digital clocks and watches have almost made reading the traditional clock obsolete. However, it is still in math textbooks, so we must teach it. I use individualized clock faces made from poster board with the hour hand colored a bright orange and the minute hand brown or another color. I show the students that the number my mouth says first is the hour and the numbers I say next are the minutes as I write (5: 30) in standard notation on the chalkboard. I ask students to place the orange hour hand on the five then move the brown minute hand slowly from its home position on the twelve to the right. As the hand passes a number, count 5-10-15-20-25-30. Now your clock is showing (5: 30), with the hour hand on the five and the minute hand on the six.

The prerequisite skill needed here is the ability to count by fives. We do this daily for as long as it takes to learn it. We chant the counting by fives orally at first using the number line and then from memory. We make a game of it. I start the chant then point to a child who takes turns as the leader and we continue until all have had a turn as leader.

We play a game I call FACE DOWN. This is really drill on showing time on a clock face. The children are divided into two teams. I call out a specific time as I write the time in standard notation on the chalkboard. The children then find that time on their clocks and turn their clock face down on their desks. I walk around the room and count the number of students who got the time correct and give one point for each correct answer to the teams.

No one knows who missed the time, since their clocks are face down on their desks. After I have checked each team for correct answers and given the teams their points, I show them what the clock face looks like when they have the right answer. We continue the game until they are no longer interested. If we are working on money at the same time, the teams are paid a certain amount of play money for each correct answer. The team with the most money is the winner, but they must first count their money correctly or forfeit the win. At first they are only paid $.05 or $.10 for each win to facilitate easy counting of the total amount of money.

COUNTING MONEY

In the beginning each student is given a packet of play money. The money can be bought from a school supply store or most variety drug stores. Most consumable math books have these manipulatives in the back of the texts. My source is from the textbooks. We first learn the value of pennies, nickels, dimes, and quarters in that order; then we learn the paper bills. We learn to distinguish one coin from another by size and engravings. The color of the copper penny makes it easy to find.

The students begin by separating and categorizing their packets of coins on their desktops. Then we begin to talk about the value of these coins. A penny equals one cent. How many cents are in a five-cent

FUN AND GAMES WITH MATH

coin or a nickel? How many one-cent coins would it

take to equal a dime or ten cents? We continue this

with other coins until we have learned their value.

MATH IS NEWS-WORTHY:

COMPARISON SHOPPING

Most any subject can be taught by using the newspaper. In this section I tell how I have used it with math lessons.

To make children more aware of a bargain, we pretend that we are going shopping. Before we go grocery shopping, we look at the ads in the newspaper. We have a list of foods and other household items to buy. I purposely try to make a list that can be found in two or more of the grocery store ads. The children work in pairs or small groups. Sometimes they work alone. The various grocery store ads are pulled from the paper as the work begins. One looks for the price of a gallon of milk at STORE A while the other two look for the same item at STORE B and STORE C. Then they proceed with the other items on the list-comparing prices. When they finish, they have the price of the items from all three stores, and they total the three columns of prices and decide which is the best buy.

We use department store ads to shop for clothes or birthday presents for family and friends. We check all ads to see where we might get the best bargains. We shop within a budget. For instance, I might say, "You have $30 to spend on a present for your mother. You cannot go over that amount. You cannot spend less than $25. Shop wisely." At different seasons we shop for different things, like Christmas presents or Easter clothes.

WEATHER

Weather charts and maps provide opportunities to correlate geography, science, and math. Geography lessons can be designed through the use of the weather map; science lessons can include understandings of fronts, pressure systems, and weather patterns. Math lessons can include activities such as the ones that follow: One team of two or three students can prepare activities for other teams to compete in games using weather maps, charts, and questions as shown in the following example.

(1) How many hours are there between sunrise and sunset?

(2) What are the high and low temperatures for today? What is the range (difference)

51

between the two?

(3) Compute the temperature range for each city in the Mid-South. How does this range compare with the range in Memphis?

(4) Which city had the greatest amount of precipitation? Did it rain in Tennessee? Where? The following two pages have weather information for these answers and others you think of.

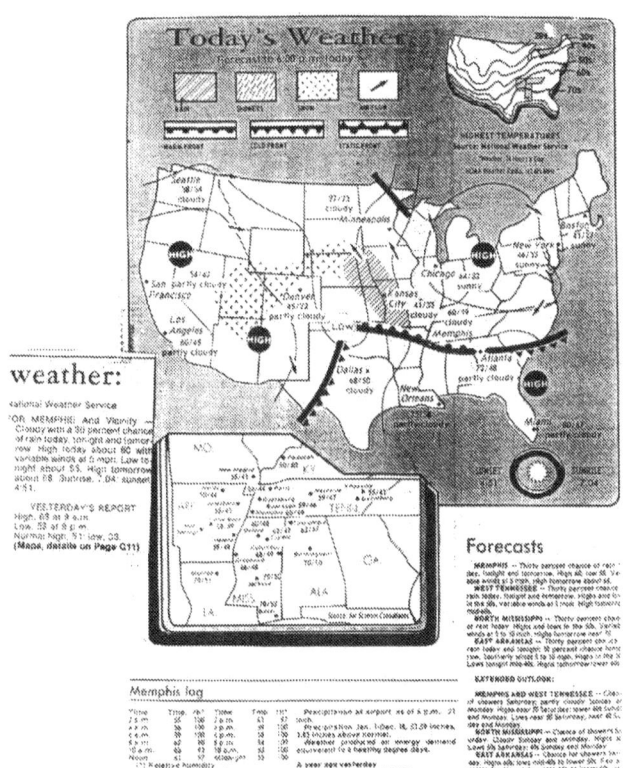

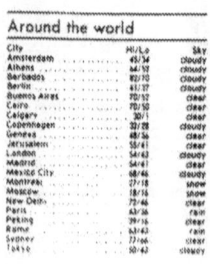

COMPARING HIGH AND LOW TEMPERATURES

This information can be collected over a period of a week, a month, three months, or more, and a daily or weekly graph can be made to show the information. The temperatures around the world can be collected and graphed in the same way. Students can then make up their own problems about the graphs. When children give words to a problem, they can usually solve it.

Problems about the graphs. When children give words to a problem, they can usually solve it.

The temperature of two cities can be compared by using > < or = symbols. A comparison can be made between two days, two weeks, or two months.

The average temperature can be calculated by the week, month, or season. This information can also be shown graphically, and story problems can be written about it.

The graph below shows the high temperatures for one week.

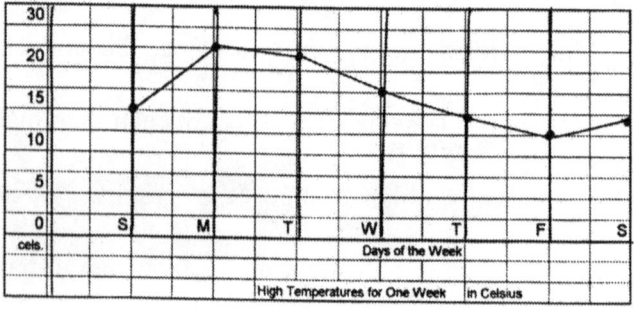

USING SPORT SCORES

Many intermediate students are motivated to do math by using the sports scores of the local high school teams as well as the professional teams' scores. The children need the sport section of the newspaper with those statistics listed.

There are many activities that can be done with this section of the paper. The students can find the averages of specific players on a designated team. The entire team's average can be computed.

Students can select a team and compute the percentage of games won and lost. They might also calculate the percentage of games won in a particular week. If they choose a basketball team, the percentage of shots attempted and made and the percentage of free shots attempted and made could be computed and graphed.

For football, the students can compute the percentage of wins for each team in the various divisions. The divisions might be ranked in order based on the won-lost records. I am sure that you can think of many other ways to use this section of the newspaper—just let your imagination wonder.

The following page is part of the sport section of a newspaper that shows scores and information that could be used to answer suggested questions and activities above.

FROM SPORTS PAGES

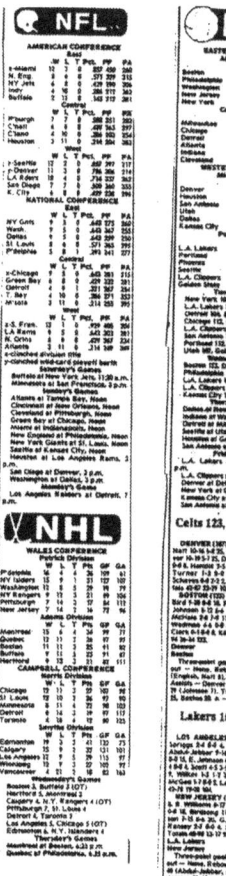

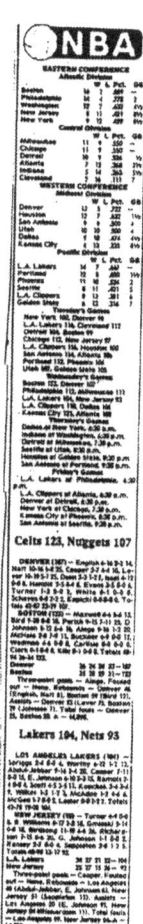

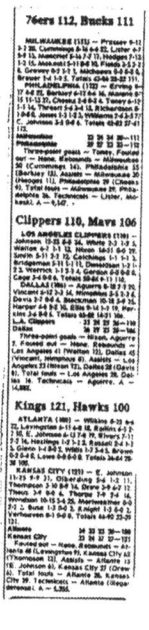

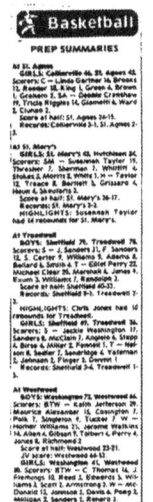

FINDING NUMBER WORDS

This activity helps reinforce the word names for numbers. Each student is given one or two pages from the newspaper to look for number words. When number words are found, they are cut out and pasted on a sheet of manila paper or notebook paper. Primary children may need some assistance in their search for number names. The objective for this assignment can be any number words or specific number words.

NUMBER WORDS—SEARCH GAME

Older elementary students enjoy the competition of this game. When the instructor says, "Go." The groups will have five minutes to find as many number words as possible. At the end of the five minutes, one student from each group will write the words they found on the chalkboard. The group that has the most words is the winner.

If vocabulary study is needed, the winning group from the word search game could direct the spelling lesson for the next few days. The entire class should be able to spell and define the words the whole class found in the newspaper. The computers in the school can be used to make a word search game or a crossword puzzle from these words and their definitions.   Among the list of words could be recession, inflation, loans, stock, price, savings, creditors, double, discount, money, bargains, etc. However, the list of words could be words the children use daily in math class.   On the next page is an example of a word search game using math terms familiar to the children.

Circle the words when you find them. The words can be found horizontally and vertically.

addend	comparison	factor
angle	cube	fraction
area	decade	numerator
average	denominator	quotient
billion	digit	rectangle
centimeter	divisor	remainder
circle	equation	zero

```
Z  E  R  O  A  A  D  D  E  N  D  B  A  R  E  A  B
C  Q  E  C  D  N  E  E  F  G  I  H  I  C  J  V  I
E  U  M  K  L  G  M  C  N  O  G  P  Q  I  R  E  L
N  A  A  S  T  L  U  A  V  M  I  L  E  R  F  R  L
T  T  I  W  X  E  Y  D  Z  A  T  B  C  C  R  A  I
I  I  N  D  E  F  G  E  H  I  J  K  L  L  A  G  O
M  O  D  E  N  O  M  I  N  A  T  O  R  E  C  E  N
E  N  E  M  D  I  V  I  S  O  R  N  O  P  T  Q  U
T  R  R  E  C  T  A  N  G  L  E  S  T  U  I  V  M
E  W  X  N  U  M  E  R  A  T  O  R  Y  Z  O  A  E
R  B  C  D  B  C  O  M  P  A  R  I  S  O  N  E  R
F  G  H  I  E  J  K  L  M  N  O  P  Q  R  S  T  A
F  A  C  T  O  R  U  Q  U  O  T  I  E  N  T  V  L
```

FINDING NUMBERS WITH SPECIFIC

PLACE VALUE

There is something about cutting and pasting in the primary grades that makes a lasting impression. When children find a numeral with two or more digits, they can read them and expand them like this: (126 is 100 + 20 + 6 or 1 hundred, 2 tens, and 6 ones. The children are given two or three pages of the newspaper to find and cut out the numerals. They cut and paste the numerals they find on a sheet of notebook paper.

First graders may sometimes need some help doing this activity. Older children can search for much larger numerals and do expanded notations like these: (1,233,142 is 1,000,000 + 200,000 + 30,000 + 3,000 + 100 + 40 + 2 or 1 million, 2 hundred thousands, 3 ten thousands, 3 thousands, 1 hundred, 4 tens, and 2 ones. This activity lends itself to small groups or individual projects. Two-digit numerals might be required of first and second grade students, three and four digits for third graders, five to eight digits for fourth through sixth graders.

A number card game can be made to develop skill in sequencing numbers. Numbers are cut from ads and pasted on poster-board or other cardboard. These cards are stored in a box. Two or more players can play the game. The numbers are laid on the playing surface face up. The object of the game is to place the cards in proper sequence from the lowest number to the highest, skipping those numbers not in the box.

USING FACTUAL INFORMATION FOR

GRAPHS

Groups of three or four students can be organized to record, chart, and graph some of the vital statistics that appear in the daily newspaper for one or more weeks. For example, one or more groups may be assigned some of the following topics. When completed, the different groups bring their data together for the class to analyze.

Topics

(1) The number of births

(2) The average age of deaths

(3) The average age and range of men and women getting married

(4) The average age and range of men and women getting divorced

(5) The number of marriages vs. the number of divorces

(6) The difference between the number of births and the number of deaths

(7) The number of births by sexes

Story problems can be made by the students about the information they have graphed. The graphs and story problems can be shared with other classrooms and will become a learning situation for them also.

READING A T.V. SCHEDULE

Reading a TV schedule is part of daily life. It is imperative that children learn to read it adequately. Schedules come in many varieties. Some are in book form and are delivered free with a subscription to cable TV, once each month. Another comes with the local Sunday newspaper and has the TV programs listed for the week. Another weekly schedule is the popular *TV GUIDE* to which one must subscribe if it to be delivered to the home. Finally, there is the daily schedule in the daily newspaper. If children learn to read one of these types well, they should be able to read most any schedule.

For practical purposes, I shall deal with the weekly schedule found in the Sunday paper. It has morning, afternoon, and evening programs charted. Reading it is similar to reading a graph. The thirty-eight or more channels received here are listed in the left column. Across the top are listed the hours and half hours these programs may be seen locally. The time is blocked off in thirty minute intervals like this: (Morning 7:00— 7:30— 8:00— 8:30). These intervals continue until 12:30 P.M. when the afternoon schedule is in effect. The programs offered are randomly spaced in these thirty-minute slots depending on how long each program lasts.

Such questions as these could be asked. At what time does *GOOD MORNING AMERICA* air on Channel 13? How long does that program stay on the air? On what channel and at what time can *SPIDER MAN* be seen?

After asking the questions, show the children how to put their left index finger on the channel number in the left-hand column and put their right index finger on the time at the top of the page. Slowly move the left finger across the page to the right until it comes to the name of the program and stop. Then move the right finger to the right, even with the name of the program. The time the program comes on is under that right finger. Following is the TV schedule used with the questions above.

MORNING SCHEDULE

	7:00	7:30	8:00	8:30	9:00	9:30	10:00	10:30	11:00	11:30	12:00	12:30
③	CBS Morning News (Cont'd)		Good Morning	I Love Lucy	$25,000 Pyramid	Press Your Luck	The Price Is Right		Young And The Restless		News	As World Turns
⑤	Today				Donahue		Wheel Of Fortune	Santa Barbara	Facts Of Life		News	Days Of Our Lives
⑩			Sesame Street		Electric Company	Educational Programming				Spaces	Educational Programming	
⑬	Good Morning America				Barnaby Jones		Trivia Trap	Family Feud	All My Children		News	Loving
㉔	Pink Panther	Popeye	Fat Albert	Super-heroes	700 Club		Jim Bakker		Jimmy Swaggart	News	20 Minute Workout	Movie
㊱	Bugs Bunny And Friends	Flintstones	Inspector Gadget	Spider-Man	Eight Is Enough		Family		Movie: 'Perfect Gentleman' ★★ (1978) Lauren Bacall, Ruth Gordon			
CBN	Inch High Priv. Eye	My Little Margie	Dobie Gillis	Bachelor Father	Age Of Destiny			Another Life	Pat Boone, USA		Ben Casey	
WTBS	Bewitched	I Love Lucy	Movie: 'The April Fools' ★★½ (1967) Jack Lemmon, Catharine Deneuve				The Catlins	All In The Family	Perry Mason		Movie: 'This Savage Land' ★★½ (1968)	
WGN	Bozo			Beverly Hillbillies	The Waltons		Big Valley		Family		News	
NASH	Amazing Facts	Porter Wagoner	Nashville Now		New Country	Fandango	You Can Be A Star	I-40 Paradise	Pickin' At Paradise	Yesteryear Nashville	Porter Wagoner
USA	Cartoons (Cont'd)		Calliope		Candid Camera	American Homemaker	Sonya		Movie: 'Sounder' ★★★★ (1972) Cicely Tyson, Paul Winfield			
WOR	Meet The Mayors	Straight Talk		News	Romper Room		Partridge Family	Bewitched	News		The Saint	
NIC	Belle And Sebastian	Today's Special	Pinwheel								Special Delivery	
ESPN	Business Times		SportsCenter		Pocket Billiards		Harness Racing		NCAA Football: North Carolina At Clemson			
DIS	Donald Duck	Pooh Corner	You And Me, Kid	Animal World	Movie: 'Hacksaw' ★★ (1973) Tab Hunter, Susan Bracken			To Be Announced	Movie: 'The Watcher In The Woods' ★★★ (1980) Bette Davis		Disney Album	
MAX	Movie: 'With A Song In My Heart' ★★★ (1952) David Wayne (Cont'd)			James Cagney	Movie: 'Kiss Me Kate' ★★★ (1953) Kathryn Grayson, Howard Keel				Movie: 'The Great Santini' ★★★ (1979) Robert Duvall, Blythe Danner			
HBO	Movie: 'Jimmy The Kid' ★★ (1982, cont'd)		The Nightmare Of Cocaine		Movie: 'The Chosen' ★★½ (1981) Maximilian Schell, Rod Steiger				Movie: 'The Premier' ★ (1979) Kathleen Quinlan, Stephen Collins			
SHO	Movie: 'The Golden Seal' ★★½ (1983) Steve Railsback, Penelope Milford				Movie: 'The Tall Target' ★★½ Dick Powell, Paula Raymond				Movie: 'Table For Five' ★★ (1983) Jon Voight, Richard Crenna			
TMC	Movie: 'The Chosen' ★★½ (1981) Maximilian Schell, Rod Steiger				Movie: 'Funny Lady' ★★ (1975) Barbra Streisand, James Caan				Movie: 'The Hand' ★★ (1981) Michael Caine, Andrea Marcovicci			

monday

MORNING

5:00 WTBS News
WGN Notre Dame Football High-lights
ESPN Business Times
DIS Good Morning Mickey!
MAX Movie ★★★½ "Tell Me A Riddle" (1980) Melvyn Douglas, Lila Kedrova. An elderly woman, unaware she is dying, reacquaints herself with her family. 'PG'
CBN Jimmy Swaggart
PTL Blackwood Brothers
USA Biznet News
LIFE Stretch With Priscilla
5:30 ③ CBS Early Morning News
WGN Faith 20
WOR NASH Jimmy Swaggart
DIS Mickey Mouse Club
HBO Robbers, Rooftops And Witches

SHO Righteous Apples
CBN Romper Room
PTL Shape Up
LIFE It Figures
6:00 ③ CBS Morning News
⑤ NBC News At Sunrise
⑩ A.M. Memphis
㉔ 20 Minute Workout
㊱ WOR 700 Club
WTBS Funtime
WGN Chicago's First Report
NIC The Adventures Of Black Beauty "The Horsebreaker"
ESPN Business Times (Repeat)
DIS Good Morning Mickey!
SHO The Mine And The Minotaur
CBN Superbook
PTL NASH Jim Bakker
USA Cartoons
LIFE Cable Health World Report
6:30 ㊱ Wake-Up Call
㉔ Great Space Coaster

WGN Muppets
NIC Lassie
DIS Mousercise
MAX Movie ★★★ "With A Song In My Heart" See 8 p.m.
HBO Movie ★★ "Jimmy The Kid" See 5 p.m.
6:35 WTBS I Dream Of Jeannie
6:45 METV A.M. Weather
7:00 SHO Movie ★★½ "The Golden Seal" See 7 p.m.
TMC Movie ★★½ "The Chosen" See 7 p.m.
PTL Real World Of Women
LIFE 80's Women
METV Sesame Street (Repeat) (C)
7:30 LIFE What Every Baby Knows
AETV Farm Day
7:45 ⑩ AETV A.M. Weather

71

COMPARING THE COST OF

DIFFERENT THEATERS

The students are given newspapers with the movie listings in them. Then I tell them that they can choose a movie from the (G) rating or (PG) category. They are then instructed to look for the prices each theater charges for a movie ticket. They are told they can take a friend along, which will double their expense. They must plan to take enough money to buy popcorn and a soft drink. When they have calculated the cost at several local theaters, they decide which movie would be the best bargain and tell why. They figure the amount of money it will cost for two tickets and refreshments. They also compute how much money it would cost if they could go to the matinee. Following are some of the listings from which they might choose

the movie and figure the cost.

MONTHLY AND YEARLY SALARIES COMPUTED BY USING THE WANT ADS

This activity is especially good with intermediate students or older children because some of them are already thinking seriously about what they want to do when they grow up. The students are given the want ad section of the newspaper. They look through the Help Wanted section and select a job in which they are interested. Then they take the salary listed, weekly or monthly, and figure how much they will receive for a year.

For instance, if yearly salaries only are listed, like $25,000 yearly, the child can compute the weekly salary by dividing by 52 weeks. One could find the monthly salary by dividing by 12 months. This information can be shared and compared by using (> <

or =) symbols and posting the information on a chart

or on the chalkboard.

FINDING MULTIPLES OF NUMBERS

The purpose of this activity is to give practical experience through student involvement in finding multiples of numbers. The students should be directed to find large numbers from ads in the newspaper and prepare two boxes (shoe boxes will do). One box will contain numbers for which multiples must be found and the other box contains multiples of all numbers in the first box. For example, the number 5, used as a stimulus, could have the following multiples 5, 10, 15, 20, 25, 30, 35, 40, 45, 50, etc.

A game can be played with the numbers in the two boxes. Two students choose a number from the box of stimulus numbers by reaching in and taking a number without looking at them. They go to the front of the room and show and tell what the number is. The other students choose a number from the multiple box and decide which number their number is a multiple of and go to that person and stand beside them. When all students have determined their number, and have stood beside the person holding the stimulus number, they explain why their number is a multiple of that number. After students gain skill in this process, a time limit can be used and speed drills practiced. This activity could be done in teams, when competing for speed.

RATIO RESEARCH

This activity aids the student in discovering through comparison the meaning of *ratio*. You will need several sets of similar pictures of different sizes, cut out, mounted, and labeled with either size or prices used like the illustration below.

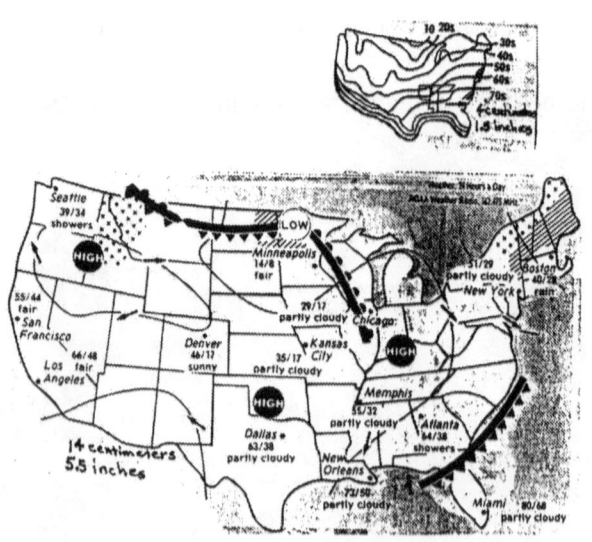

The following questions might be used with pictures like those on the previous page.

1. The smaller weather map is how many inches shorter than the larger map?

2. The larger map is how many centimeters longer than the smaller map?

3. The ratio of the smaller map to the larger map is 1.5 to 5.5.

4. The ratio, in centimeters, of the larger map to the smaller map is 14 to 4 or 14/4.

INVESTING IN STOCK

This helps the student understand the stock market and how it affects our daily life.

The students find the section of the paper where the stock reports are listed. I tell them to pretend that they have $1,000 to invest in local stock. They choose one company in which they will invest their money. Each day they look in the paper to keep track of their stock. Cross-sectioned manila paper is used to show the rise or fall of their stock on the market. Graph paper can also be used but it costs a little more.

Children can be asked to respond to questions like these.

(a) Which stocks were the most active for the day?

(b) How much would 50, 100. or 200 shares of your stock cost?

(c) Does your stock pay a dividend? How much?

(d) How many shares of your stock were sold today?

If students make money on their stock, they can purchase other stock and graph the progress of it. Following are examples of Area Stocks from the newspaper that I use with my students. They seem more interested in the Stock Market if the stocks they pretend to purchase are local. They can keep track of them in the newspaper and on local TV.

AREA STOCK

Nasdaq Stock Market

	Sales	Bid	Ask	Close	Chg
Bck Yrd Brgrs...	84	4¼	4⅝	4¼	0
BancorpSouth ..	13	34¼	35¾	34¼	-½
Boatmen's ...	1752	30¼	30⅜	30.31	-.19
Catherine's...	8243	7⅝	8	7¾	0
Chromcraft......	3	21	21¾	21½	-¼
CMC Industries .	34	3⅞	4¼	4⅜	+.14
Comm. Bnc......	0	16¾	17¼	17	0
Concord EFS ..	505	26¾	27¼	26¾	+½
Coors	654	16¼	16⅜	16.31	-.19
DeltaPine	101	21½	21⅞	21½	+⅛
Deposit Grty. ..	159	32	32¼	32½	+⅛
Equity Inns	587	10⅛	10⅜	10½	0
1st Amer.	325	33¼	33½	33.31	-.06
1st Comm.	3	24⅛	24⅜	24⅜	-⅛
1st Tenn......	1247	42½	43	42½	+¼
Fred's	12	9¾	10½	9¾	0
Gibson Grtg....	163	9⅝	9⅝	9⅝	+⅛
Insituform	624	12	12¼	12.06	-.31
Leader Fin	116	26⅝	26⅞	26¾	-⅜
Mark VII........	15	13⅝	14⅛	13⅝	-¼
Midland Fin ...	491	18	18¾	18⅛	+⅛
MSCarriers....	373	23	23½	23	0
NCBC	329	25	25½	25	0
Natl.Pic.&Frm...	10	9¾	10¼	9¾	-½
NPC Intl.	40	5⅞	6⅛	6⅛	+⅜
NW Air.......	2971	23¾	24	23¾	-⅛
NSA Int'l........	6	3¾	4¼	3⅞	0
Omega Health ...	0	4¾	5½	4⅝	0
Proffitts.......	475	22	22¾	22	0
RFS Hotel Inv. .	112	13⅞	14¼	13⅞	0
Steinmart ...	12603	9⅝	9⅝	9⅝	0
TBC..........	49	10	10⅛	10⅛	+⅛
TPI...........	211	5⅝	5⅝	5⅝	0
Tyson Foods .	1297	24⅜	24⅝	24⅝	+⅛
Union Plntr pf....	2	31	31¾	31	-⅜
Varsity Spirit....	63	13	13¾	13¾	+⅜
Wall St. Deli ...	294	8½	9¼	8½	-1

Sales in hundreds. NA — not available; w — when issued; x — ex-dividend; y — 2-for-1 split.

New York Stock Exchange

	Sales	High	Low	Close	Chg.
AMI	9	25½	25½	25½	0
AT&T.......	34225	52¼	51	52¼	+1½
Arcadian LP ...	118	24⅝	24½	24½	0
AutoZone	1286	25⅝	25	25½	+⅜
Bass PLC	2	15⅞	15⅞	15⅞	+⅜
BellSouth	6725	59	58⅛	58¾	+⅜
Boyd Gaming..	112	13⅝	13⅜	13⅝	-⅛
Dial Corp.....	1667	25⅞	25½	25⅞	+⅜
Dillard	6356	26¼	25⅝	25⅝	-¼
Dover	1496	60½	59½	59¾	+⅜
Du Pont.....	12453	56⅜	55½	56⅛	+⅝
Dyersburg......	25	5¾	5⅝	5¾	+⅛
Fed. Expr.....	2271	66⅞	65¾	66¾	+1⅜
Federated....	7325	21⅝	21⅛	21¼	-⅛
Fleming Co. ...	516	20⅞	20⅝	20⅝	+⅛
Goodyear	7358	36⅛	35¼	35¼	-⅛
Hanc'k Fab. ...	772	11⅛	10⅞	10⅞	-⅜
Int'l. Paper ...	4472	72⅝	71⅝	72½	-⅛
Jostens.......	870	21	20⅝	20¾	0
Kellogg	6943	58⅛	57	58	+1⅛
Kroger Co. ...	3607	27⅛	26⅝	26⅞	-⅜
MAPCO.......	209	54¼	53⅞	54⅛	+⅛
Maybelline	465	20⅝	20¼	20½	+⅛
MidAmer Apts ..	59	25⅜	25¼	25⅛	-⅛
Morg. Keeg.	37	15¼	15⅛	15⅛	+⅛
NationsB'k ...	5912	50⅛	49⅝	50⅛	+⅝
Nike.........	1545	77⅜	75⅞	75⅞	-1½
Perkins Fam. ..	123	12⅞	12⅛	12⅝	+½
Philip Mor. ..	22436	64	63⅜	63¾	+½
Promus Co. ..	3914	35⅝	34⅞	35¼	+⅛
Sara Lee	19123	27⅝	26⅞	27¾	+½
Schg-Plough .	9899	77¼	76⅛	76½	-½
E.W. Scripps ..	133	29	28¼	28½	-½
Sears.......	11176	51⅛	50⅝	51⅛	+⅜
Service Mstr. ..	365	24¼	24	24⅛	+⅛
SofamrDank ..	3601	25	24⅛	24¾	+¼
Storage USA ..	661	28⅜	28⅛	28⅝	+⅜
Thos. Betts	324	66½	66	66	0
Union Pltrs.....	112	23⅞	23¾	23¾	-⅛
Wal-Mart	25982	25	24⅝	24⅞	+⅛

WHEN ALL ELSE FAILS, USE ART

The greater than and less than symbol take on much more meaning when I tell the students to think of them as hungry Pac Man. Below is an example.

I place my hands together as if they were hinged and open and close them, pretending to eat anything in sight like a hungry Pac Man. I have small groups of children stand in two groups with more children in one group than in the other, and I pretend to gobble up the largest group with my hands. After this introduction, we compare abstract numbers on the chalkboard and, they draw pictures like this: $(7 > 5)$. They write on the board and say, "Seven is greater than five." I write $(5 > 3)$ on the board, and in unison the children say, "Five is greater than three." If they are allowed to draw the symbols, they can learn to write the math sentences much faster. They need lots of practice in saying correctly what they have written.

PICTURE GEOMETRIC SHAPES

These can be precut before teaching the lesson, and students can assemble them, or they could be done entirely by the instructor for a display on a bulletin board. When displayed, these shapes become a constant reminder of the geometric shape and its name. Below are some examples of the four major geo shapes that primary children are expected to know. For intermediate children, other shapes can be pictured. They can construct their own pictures and explain how they made their geodesic forms. They could be assigned certain geo shapes to use in designing a building or some other project similar to the illustrations that follow.

GEOMETRIC SHAPES

Ideas for Bulletin board

SPRING FLOWERS MADE FROM

GEOMETRIC SHAPES

An iris can be made from four circles, one rectangle, two triangles, and a square piece of construction paper on which to mount the flower. Since irises come in many colors, the circles can be made from a variety of colored construction sheets. The rectangles and triangles should be green, because they make the stem and the leaves. The square pieces of paper can be a contrasting color to emphasize the petals of the flower.

A graph can be made which shows how many flowers of each color were made. Questions or story problems can then be written about the graph. These beautiful little flowers can be hung on the lockers or displayed some other place to decorate the classroom.

An example of the iris is shown below.

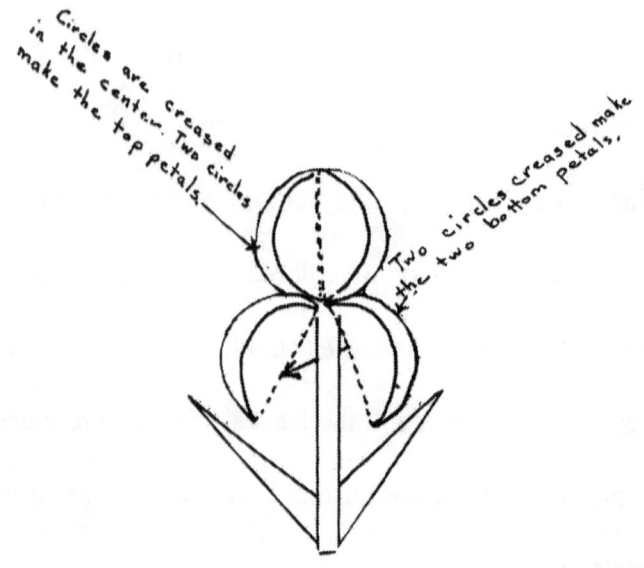

GEOMETRIC SHAPES MAKE LOVELY

CHRISTMAS ORNAMENTS

The circle, square, and triangle make pretty Christmas tree ornaments. Other geo shapes do also, but I like these best. Six circles, cut the same size, are needed for the circle ornament. Five squares, cut the same size, are needed for the square ornament. Four triangles cut from the same pattern are needed to make the triangular ornament. These shapes can be cut from colored construction paper or a heavy-weight Christmas wrapping paper.

Instructions For Making Ornaments

Leave one circle flat while the other five are folded in half and creased. Then cut the five folded pieces from the fold in the center toward the outside edge of the half circle to within two centimeters of the edge. One of each of the other shapes is to be left flat and whole. Four squares and three triangles are to be folded in half and cut from the fold toward the outside edge.

The illustration below will explain how to fold and cut the geo shapes.

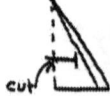 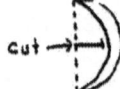 

To assemble, hold the folded slit pieces in one hand and the flat circle in the other hand. Slowly scoot the folded piece on the circle and gently spread it apart. Do this with the other four cut pieces and a perfect five-point star appears on the flat circle. You have made a round three dimensional circular ornament. Each folded cut piece of the squares and the triangles is to be placed on the flat geo shape in the same manner but do these on each corner. The square makes a four-point star and the triangle makes a three-point star. All ornaments are three dimensional.

When the ornaments have been assembled, make a graph that shows how many of each kind of ornament was made. The students can write questions or story problems about the graph and take turns answering each other's questions.

We display our handy work on a hand-made tree. Using green construction paper, each child traces the shape of his hand and cuts out the handprint. The fingers are slightly curled to look as if they are leaves of an evergreen tree. The hand shapes are then taped to a flat surface, like a wall, in the shape of a Christmas tree. The ornaments are hung from the tree with hooks made from open paper clips. See picture below.

CUTTING AND PASTING ACTIVITIES

A small group or an entire class can prepare a bulletin board using geometric shapes found in a three to five-day period in the newspaper. Colored picture collages can be made from pictures cut from magazines or newspapers to show: a point, a segment, a ray, a solid, a plane, a circle, a square, a sphere, a cone, a triangle, etc.

Young primary students can find pictures of houses and cut them out. They can then look for numbers and write their house numbers on these pictures of houses. Colored construction paper can be used to cut the shape of a house, and children can paste house numbers on these if pictures cannot be found.

Another activity for early primary students is to use construction paper and cut out the shape of a telephone. Then they can look in the newspaper and find numbers to cut and paste their telephone numbers on the construction paper phones. These make an interesting bulletin board display.

MATH FLOWERS FROM PAPER

PLATES

Math problems can be written on petals cut from paper plates. Slits that fit the petals are cut in another plate so petals can be inserted and a flower made. The center and the petals can be as colorful as your imagination allows. For four players, four plates with slits are needed for the centers of the flowers, and as many petals as needed to form the four flowers should be cut.

The problems are written on the front of the petals, and the answers are written on the back for self-checking. To play the game, the petals are laid on a flat surface face up. Each player is given a plate or center of the flower. The players take turns drawing a petal and trying to answer the problem written on it. If they answer correctly, they insert the petal in the flower. If the player cannot answer the problem, the petal is laid aside. The first one to complete a flower is the winner.

The problems can address any skill that needs to be practiced. If numeration is needed, the problem might be this: (1, 2, 3, —, 5, —, —, 8, 9. —). If multiplication facts need attention, the problem could be this: (3 x 9 =?).

The illustration below will explain how the math flowers are made.

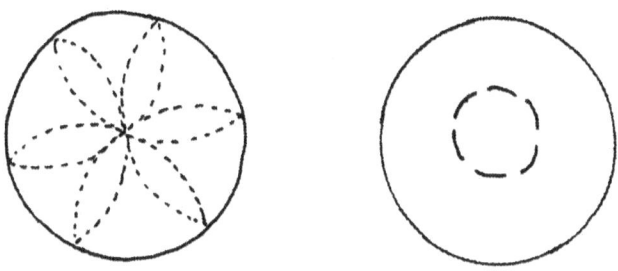

Petals cut from one plate fit the slots cut in another

paper plate.

CHILDREN CAN DRAW PICTURES OF

HOLIDAY SCENES

This activity is a pleasant diversion from the story problems in the textbook. For instance, have the children draw Halloween pictures. Show the pictures to the class. Have everyone write one story problem about the picture or children can simply tell the problems. Show another picture and do the same until about five or six pictures have been shown. Continue to show a few pictures each day until all have been shown and questions have been written or asked orally about them.

These pictures with their story problems can be made into a booklet and used again. The same type of activity can be applied to most other holidays. Some of my students' pictures and story problems are included on the following pages as examples. These pictures were made by some third grade students. Children love this activity.

HALLOWEEN STORY PROBLEM AND

PICTURE

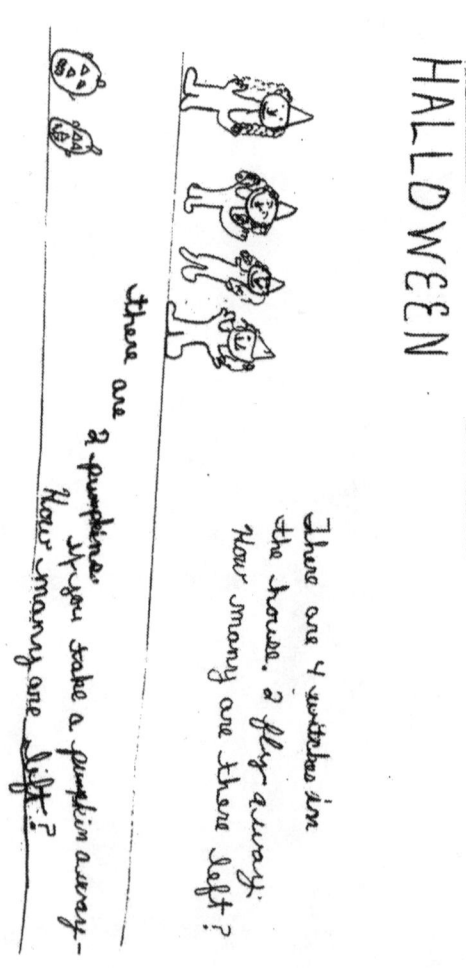

HALLOWEEN

There are 4 witches in the house. 2 fly away. How many are there left?

There are 2 pumpkins. If you take a pumpkin away— How many are left?

HAPPY HALLOWEEN STORY AND

PICTURE

HAPPY Halloween

There are 2 ghost
and 3 Pumpkins.
How many in all?

There is 1 bat
and 3 monsters.
How many in all? ____

IT IS LESS OF A DILLY TO TEACH

MATH DAILY

Activities using grocery store items are numerous. Children can recognize and read the prices of items. They can add the number of ounces or millimeters in a designated number of canned goods. The child can look at various packages and identify the ones containing more or less bulk weight. Intermediate children can figure out the unit price of items. Such questions as these might be asked. If tomato sauce is 3/$1, or bell peppers are 4/$1, what is the cost of one can of tomato sauce and the cost of one bell pepper?

The greater than and less than symbols take on more meaning when the child looks at two cans of food and compares their size. You say, "You know which can has the most food inside by looking at the size of the two cans. Let's see what that would look like written as a math sentence." Write: (18 oz. > 14 oz.). You can think of many more comparisons and problems, just use your imagination.

SHOPPING WITH COUPONS

Have the child figure the cost of specific items minus coupons. If the store doubles the value of the coupon, that is another math problem to solve. The child can multiply the value of the coupon by two and then subtract the amount from the list price. I cut coupons out of the weekly and daily newspapers and current magazines to use with my classes. Most people have access to trade coupons, too. Questions like this could be asked. If bread cost $1.49 for a loaf, what will it cost if I use this coupon worth $.50? What would it cost if the coupon's value was doubled?

The child can figure the amount of change received when they buy an item like the loaf of bread. Say, "If I hand the clerk $2.00 to pay for the bread and give her the coupon, what change will I get back?"

Have the child tell the coins needed to buy a particular item. In my classroom, I use a play store for such activities. For instance, I have an item priced at $.49. Then I tell the child I want seven coins to pay for that item. The child should give me one quarter, two dimes, and four pennies. You can think of many more questions I am sure as you do your shopping.

ACTIVITIES USING YOUR CAR

The child can write down the mileage at the beginning and end of each day. Then the distance the car has been driven in one day can be calculated. If this mileage is kept for a week then the child could figure how many miles the car is driven in that length of time.

The odometer can be used for place value with the tenths being in ones place and the tens, hundreds, thousands, ten thousands, etc., being in their own place. A question like this could be asked: "Read me the mileage before we drive to school. How many miles did we drive to get to school? What is the mileage on the odometer now?

How much more do you pay for full service than for self-service? The prices will vary across the country. Students will probably find that it is cheaper to pump their gas themselves at a self- service station.

Have children practice multiplying by computing the cost of a specific number of gallons or liters of gasoline. Have them do this for self-service and for full service.

"WORDPROBLEM" IS NOT A DIRTY

WORD

Colorful folders to keep word problems in are sometimes self-motivators. Colored Duo-tang folders with pictures cut and pasted on the outside may invite a child to look inside. Folders with Disney characters, Care Bears, Aladdin, Garfield, and Super Heroes can be purchased, but it is less expensive to decorate them yourself.

Make up word problems from everyday life to put in these folders. Such as these:

(1) There are 5 apples, 2 grapefruits, 4 oranges in the

fruit bowl. How many pieces of fruit

are there in all?

(2) We bought 6 twelve-ounce cans of Coke. John, Fred, and I each drank one a piece. How many Cokes are left?

(3) If oranges are 10 for $1.00, what is the cost of each orange?

(4) Grapefruits are $.25 each. How much would 6 grapefruits cost?

Make Up Zany Problems

(5) Mr. Mac-E-Doodle had 4 gruffy goats and 5 frolicking cats. How many animals did

he has in all?

(6) Mr. Hackelbaker has 15 pet lizards. His wife found 5 of the ugliest lizards you ever did

see and gave them to Mr. Hackelbaker. How many lizards does he have now?

(7) There are 46 steps in the old round tower. Fritter

Fang climbed 24 of them. How many

more steps does he have to climb?

(8) Magic Min knows 69 magic tricks. She used 35 of

them in her magic show. How many

tricks were not used?

INDIVIDUALIZE PROBLEMS

Problems like those in the child's math text can be individualized by substituting the names of the children in place of those in the text. Problems become much more meaningful for John and Jan when you transpose a problem from the book like this. John bought Jan a birthday card for $.50 and a present for $2.28. How much did John pay in all? Another example might be this. Jan's mother gave her $5.00 to buy these items: bread $.91, a gallon of milk $2.49. Will Jan have enough money? How much change will she get? I am sure you can think of many more problems.

KEY WORDS IN THE ASKING

SENTENCE

I tell my students that the sentence near the end of a word problem is usually the asking sentence. It is the sentence that asks something about the facts in the story. It always ends with a (?) question mark. Certain words in that sentence tell exactly what to do with the facts presented in the story problem. We do lots of practice just looking for the asking sentence. Then we begin practice drills in finding the key words.

A poster made with flowers will be a constant reminder of key words. The centers of the flowers can have the (+), (x), (-), (-:-) symbols on them. The petals of the flowers can have the key words written on them. An example of the poster is on the following page.

After constructing the poster, story problems can be cut from old math workbooks or from dittos left over or problems made by the students and teachers. These problems can be glued to cards for use in practice drills. The key words can be highlighted with a yellow marker in the beginning to teach the words and the process they indicate.

PLAY "MAGIC FLOWER GARDEN"

Use the story problem cards made with the key words highlighted in yellow for the game. The cards can be dealt out four or five face down on the desk to each player. Taking turns, each player picks a card and reads the story either orally or silently. Since the key words are highlighted; they are easily found by the child. The child looks closely at the Magic Flower Garden and locates the process indicated by the key word and proceeds to solve the problem.

Two points are scored each time. One point is given for finding the right procedure to solve the problem, and the other one is for solving it correctly. An individual can be the winner, or it can be a team that wins.

Play money can be given, $.25 for each point earned or more money if one so chooses. The individual or the team with the most money is the winner. Students like getting money, and it gives them practice in counting money. Regardless of whether points or money are given for correct answers, the child enjoys having fun while learning.

The Key Words poster idea is illustrated for your observation and use with this particular game. Illustrations are on the following pages.

KEY WORD POSTER IDEA

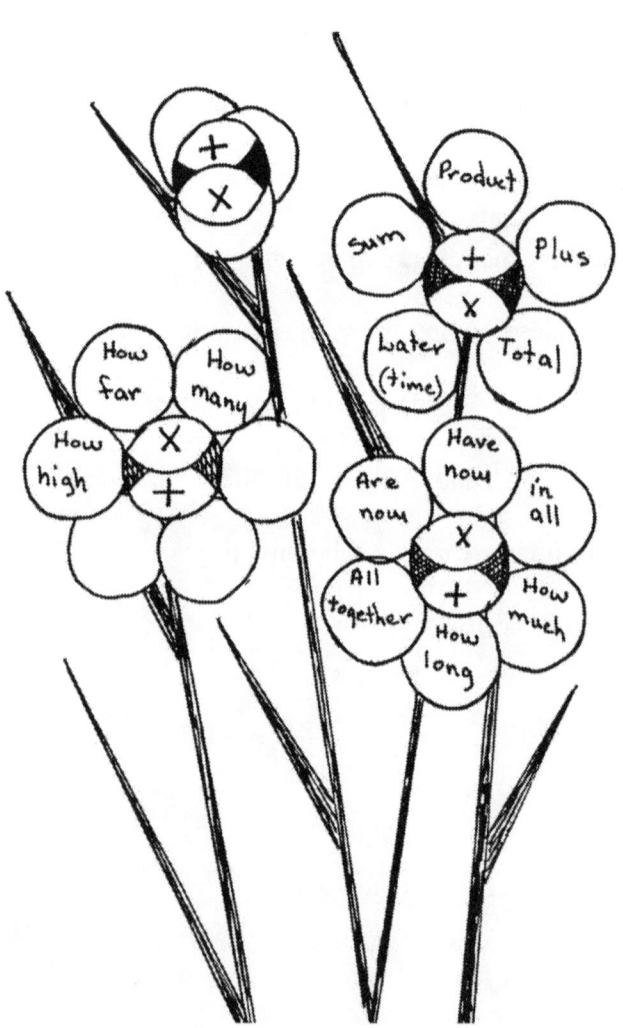

KEY WORD POSTER IDEA

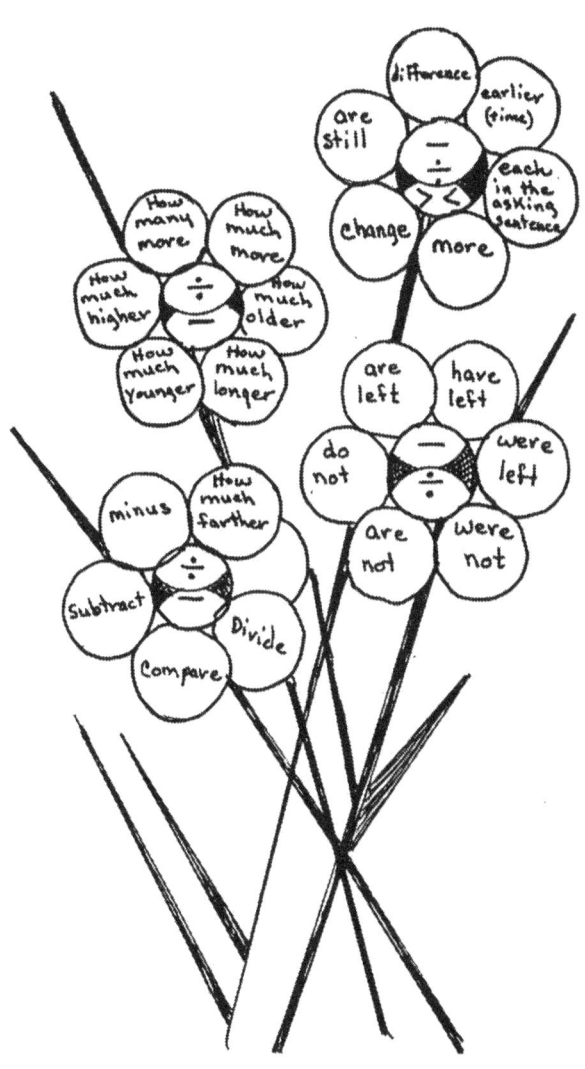

117

LITERARY APPROACHES TO TEACHING MATH

Word Cinquain—A Five-Line Poem

How can math possibly be a subject for creative writing? It can and does present many outlets for creative expression. Word cinquains are an easy, almost magical, way to write poems. Anyone can be a poet using this technique. Following is the recipe for this magic potion.

1st. line—noun (subject)—one word

2nd. line—adjectives (describe subject)—two words

3rd. line—verbs (subject's action)—three words

4th. line—adverbs or (feelings about subject)—four words

5th. line—synonym (for subject of poem)—one word

Following are some examples of word cinquains written by my math students. Steve and Angie were attending remedial reading and math classes when they wrote these poems.

Math

numbers, problems

add, subtract, multiply

regrouping is the best

arithmetic

by Steve

Multiplication

adding, regrouping

thinking, learning, figuring

I like adding fast

product

by Angie

Division

separable, splittable

multiply, subtract, divide

quick way to subtract

separation

by Jim

Subtraction

larger, smaller

withdraw, take away

Regrouping is more difficult.

reduction

by Jan

USING WORD SEARCH AND

CROSSWORDS

The teaching of the math vocabulary need not be drudgery. It can be done in a meaningful but interesting way. Word search games help students to recognize and familiarize themselves with the words to be learned. At first the teacher can make the word search games. This can be done by writing the words on graph paper (one letter per space), then filling in any blank square spaces with randomly selected letters of the alphabet. If you have the software and a computer, the job can be done in less than half the time required to make them with graph paper. The students can be taught to make these word search games by either of the methods described above.

On the following page is an example of a word search game using some math vocabulary words.

MATH VOCABULARY WORD SEARCH

See how many of the vocabulary words you can find in this word search game. The words run across and down.

base ten	length
century	mile
diagonal	millennium
estimate	multiple
foot	numeral
gallon	ounce
geometry	pint
grouping	perimeter
hexagon	pentagon
hypotenuse	pound
inch	product

segment

set

ton

two

unit

volume

yard

zero

```
B  A  S  E  T  E  N  C  E  N  T  U  R  Y  A
E  S  T  I  M  A  T  E  F  O  O  T  G  A  B
G  H  E  X  A  G  O  N  I  N  C  H  A  R  V
E  Y  P  O  U  N  D  H  T  G  N  E  L  D  O
O  P  I  N  T  W  O  A  O  B  C  D  L  I  L
M  O  U  N  C  E  E  F  N  Z  E  R  O  A  U
E  T  S  E  T  G  G  R  O  U  P  I  N  G  M
T  E  M  I  L  L  E  N  N  I  U  M  M  O  E
R  N  U  M  E  R  A  L  U  N  I  T  I  N  H
Y  U  P  E  R  I  M  E  T  E  R  R  L  A  I
J  S  P  E  N  T  A  G  O  N  K  L  E  L  N
O  E  L  P  I  T  L  U  M  M  P  Q  E  L  S
P  R  O  D  U  C  T  S  E  G  M  E  N  T  V
```

CROSSWORDS OR ACROSTICS

Cross-word puzzles can be constructed by both students and teacher. The overhead projector is invaluable for introducing the making of this kind of puzzle. We work in small groups. The groups start by writing definitions to the words we are using in the puzzle. Then they try to shorten definitions to use for clues. If they are doing this on graph paper, they write their clues on notebook paper and use the graph paper to fit their words inside. If the appropriate software is available to use on a computer, this puzzle making is done in a jiffy by following simple instructions that come with the software.

Some student examples follow.

NUMBER WORD ACROSTIC

by Pam

1. Ten - eight =

2. Eight - four =

3. Five + three =

4. Ten - six =

5. Five - three =

6. Four - three =

7. Sixteen - six=

8. Two + one =

9. Ten - nine =

10. Zero + one =

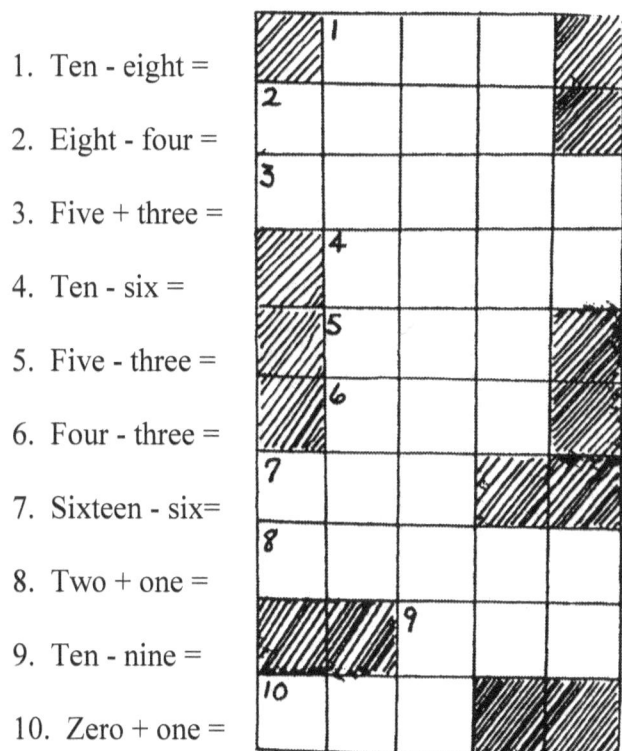

The four letters in the last vertical column will spell a familiar word if you answered correctly.

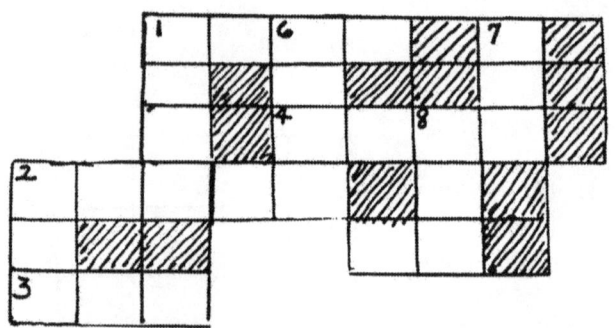

NUMBER WORDS AND NOUNS

by Eric, Jim, Keith, and Scott

Across: 1. 4, —, 6, 7, 8

2. 1, 2, —, 4, 5

3. 0, —, 2, 3, 4

4. A bird lays eggs in a —.

5. You use an — to chop a tree.

Down: 1. 3, —, 5, 6

2. 1, 2, —, 4, 5

6. Tarzan swings on a —.

7. You lift a fish in a boat with a —.

8. 1, 2, 3, 4, 5, —, 7

1.									
		2.							
			3.						
	4.								
		5.							
			6.						
8.									
			9.						
7. M	A	T	H	I	S	F	U	N	
			10.						

MATH ACROSTIC

by Sam, Frank, Tony, and Bruce

CLUES:

1. To say one, two, three, is to - - - - -.

2. A number that comes after two - - - - -.

129

3. If a number is not odd, it is - - - -.

4. This number comes after four, - - - -.

5. S- - - - is a number.

6. This number comes after five, - - -.

8. F - - r, is a number.

9. A number that comes before nine is - - - - -.

10. If your answer is not wrong, it is - - - - -.

CONCLUSION

This book is meant to be an idea starter for teachers and parents. All of these activities have been used successfully in my own classroom. They work like magic for me. The games and activities are inexpensive and easily accessible to both parents and teachers. I sincerely hope you enjoy using this book.

ABOUT THE AUTHOR

Ouida Simmons is a retired Elementary Teacher and Guidance Counselor. She worked in the Memphis City Schools for 28 years. Mrs. Simmons had three children. Her oldest son died in July 2000 from complications of heart surgery and a heart attack. Her youngest son died from a heart attack on June 1, 2003. She has 6 grandchildren and 5 great grandchildren.

Mrs. Simmons has done many things in the field of writing. As a teacher she taught other teachers "Writing Across the Curriculum". This involved lecturing across the state to educators. She has attended many Writers Conferences and has lectured at some. She has entered many literary contests for writing in nonfiction, fiction, essays and poetry and has won many first place awards for her efforts. She has given presentations on local educational TV Stations. She now teaches a Creative Writing Class at a Senior Citizen Center. Occasionally she still teaches Creative Writing at a local Community College. Mrs. Simmons is the past president, for 2 years, of the Mid South Writers Association. While serving as president of the writer's group, she worked on the editorial staff of their publication, *Writers on the River*.

She is a member of several local, state and national organizations. These organizations include: Mid South Writers Association, Poetry Society of Tennessee, National League of American Pen Women and The Tennessee Writers Alliance. She has been published in several literary publications

Mrs. Simmons was born on January 1, 1930 in Sardis, TN. At the present time, she lives at home with her new husband, Doc Nichols, in Lepanto, Arkansas. She and Doc love to travel and have done so extensively since her retirement from the Memphis City School System.